W9-CJP-739

Northern Lights

by Grace Hansen

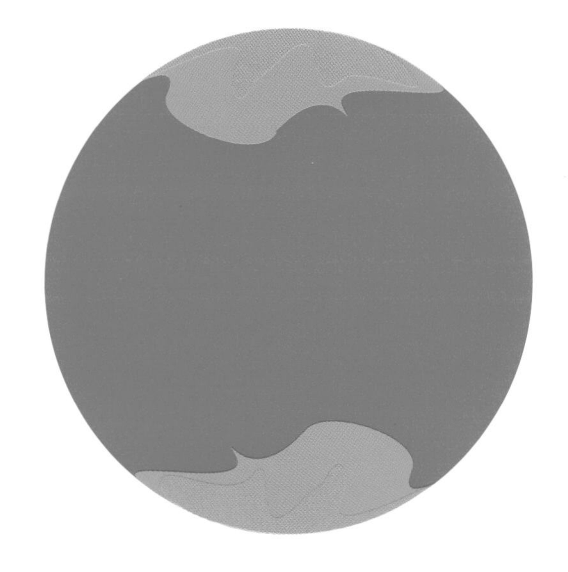

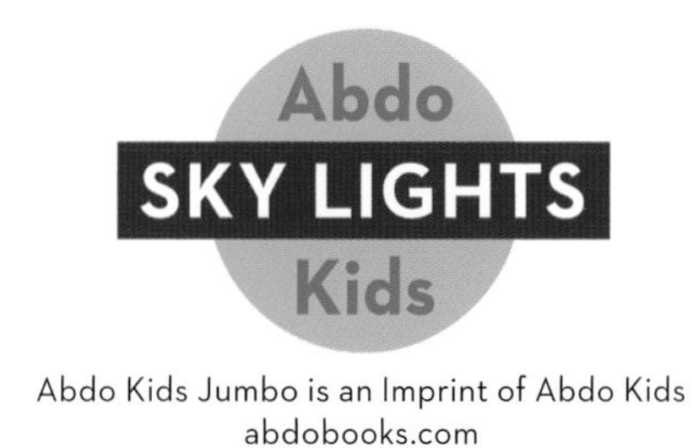

Abdo Kids Jumbo is an Imprint of Abdo Kids
abdobooks.com

abdobooks.com

Published by Abdo Kids, a division of ABDO, P.O. Box 398166, Minneapolis, Minnesota 55439.

Abdo Kids Jumbo™ is a trademark and logo of Abdo Kids.

Printed in the United States of America, North Mankato, Minnesota.

102019

012020

Photo Credits: iStock, Science Source, Shutterstock

Production Contributors: Teddy Borth, Jennie Forsberg, Grace Hansen
Design Contributors: Dorothy Toth, Pakou Moua

Library of Congress Control Number: 2019941285

Publisher's Cataloging-in-Publication Data

Names: Hansen, Grace, author.

Title: Northern lights / by Grace Hansen

Description: Minneapolis, Minnesota : Abdo Kids, 2020 | Series: Sky lights | Includes online resources and index.

Identifiers: ISBN 9781532189098 (lib. bdg.) | ISBN 9781532189586 (ebook) | ISBN 9781098200565 (Read-to-Me ebook)

Subjects: LCSH: Auroras--Juvenile literature. | Northern lights--Juvenile literature. | Geomagnetism--Juvenile literature. | Colors--Juvenile literature. | Light--Juvenile literature. | Solar wind--Juvenile literature.

Classification: DDC 538.768--dc23

Table of Contents

Auroras in the Sky

The Northern Lights look like magic. But they are real! And science can explain why they happen.

The Science Behind Auroras

The sun is the center of our galaxy. Everything in our galaxy orbits the sun. It is very big and very hot.

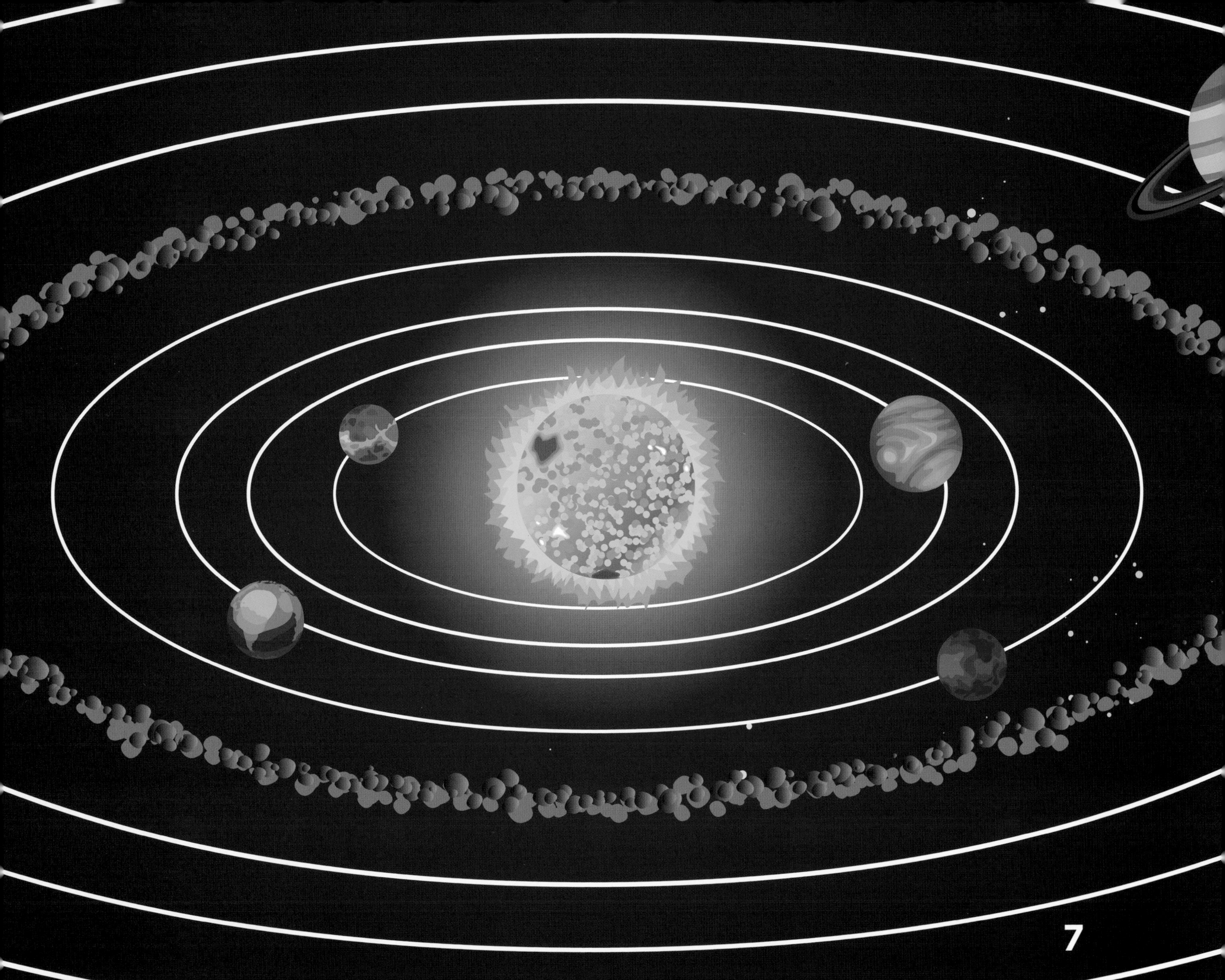

Every second of every day, the sun shoots **charged particles** into space. **Solar winds** push the particles toward Earth.

Earth is a giant magnet. It has a magnetic field that surrounds it. The magnetic field stretches out into space. It protects us from most of the **charged particles** the sun fires off.

sun

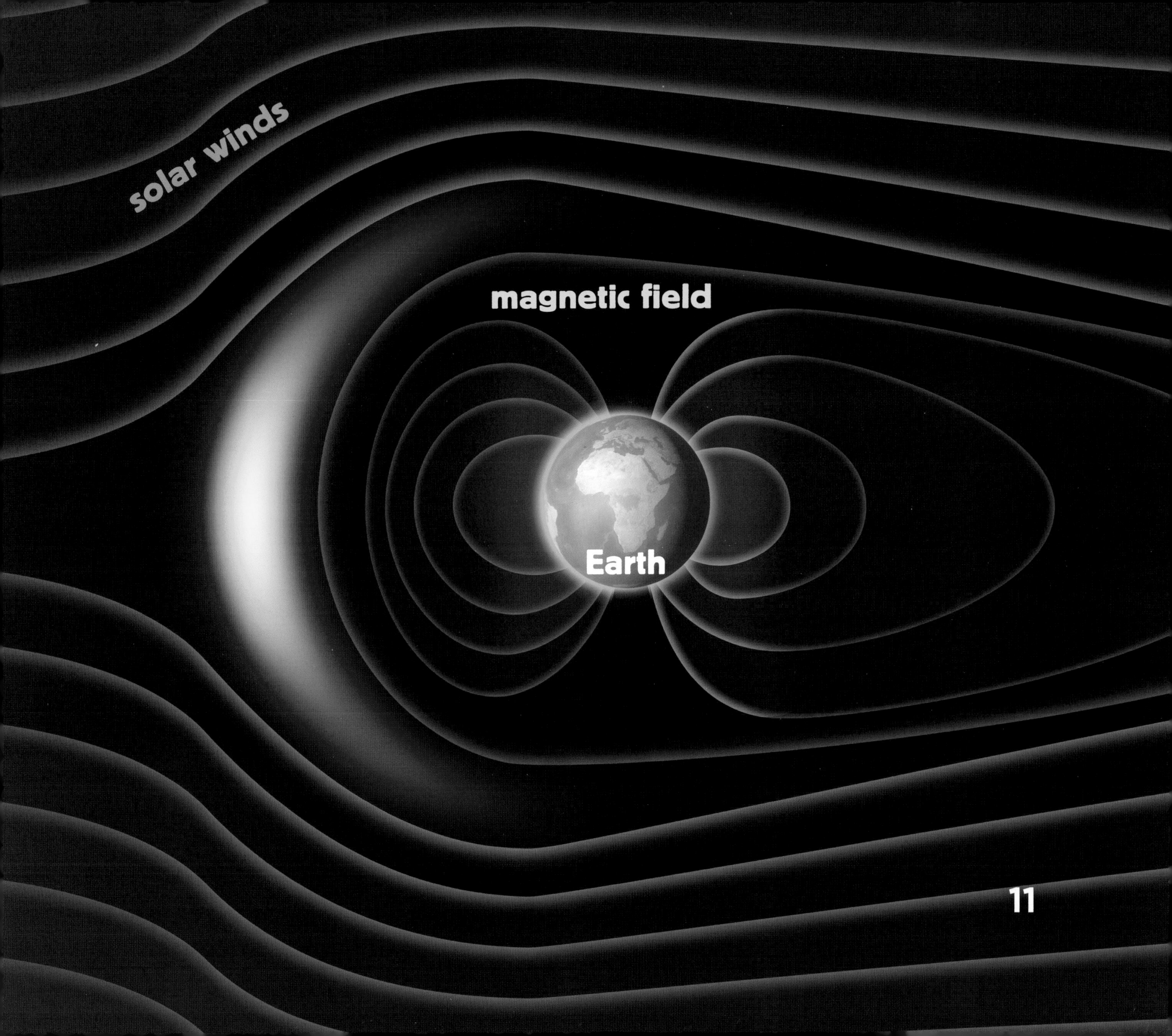

solar winds

magnetic field

Earth

However, some of these particles make it into Earth's **atmosphere**. The thermosphere is the second to last layer. There are nitrogen and oxygen **molecules** in the thermosphere.

The **charged particles** crash into nitrogen and oxygen **molecules**. Energy from the particles transfers to the molecules. This excites the molecules.

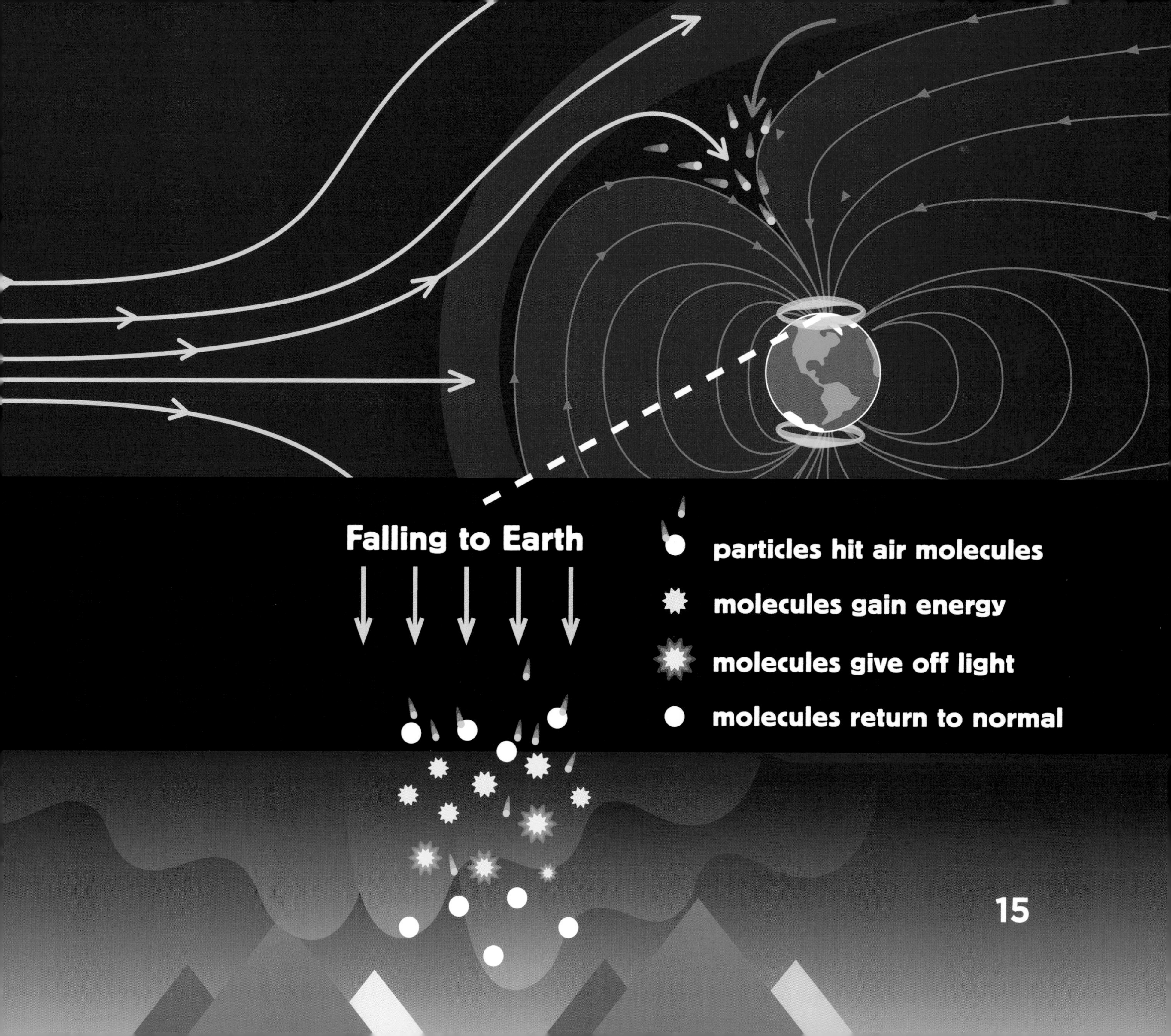
Falling to Earth
particles hit air molecules
molecules gain energy
molecules give off light
molecules return to normal

To calm down, the **molecules** release the energy. We see this energy as light. Oxygen gives off greenish or red light. Nitrogen often **emits** blue light.

Where to See Auroras

Like all magnets, Earth's magnetic pull is strongest at the North and South Poles. The magnetic field lines come together at the poles. This is why you are more likely to see the auroras near the poles.

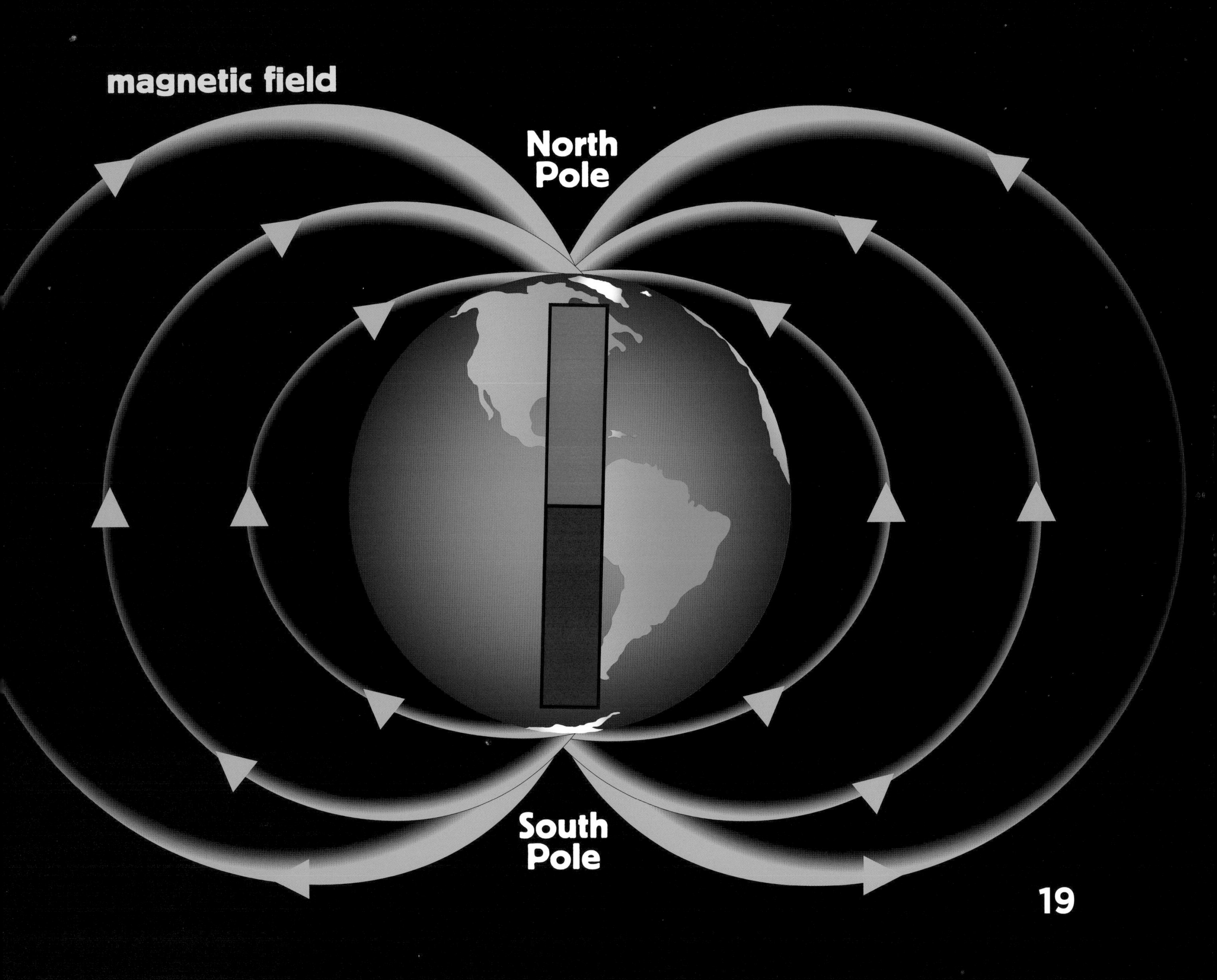
magnetic field
North Pole
South Pole

The International Space Station (ISS) orbits Earth in the thermosphere. On Earth, we see the auroras from below. On the ISS, astronauts can see the auroras from above!

More Facts

- In the northern hemisphere, the auroras are called aurora borealis. *Aurora* is the name of the Roman goddess of dawn. The Greek name for wind in the north is *Boreas*.

- Earth is not the only planet that has auroras. Jupiter, Saturn, Uranus, and Neptune each have auroras too.

- There are cave paintings in France that may show the Northern Lights. The paintings date back 30,000 years.

Glossary

atmosphere – the gases surrounding Earth.

charged particle – a particle with an electric charge.

emit – to give off.

galaxy – a collection of billions of stars, planets, and other matter held together by gravity. Earth and the sun are a part of the Milky Way galaxy.

molecule – the smallest unit of a substance that has all the properties of that substance. A molecule is made up of a single atom or group of atoms.

solar wind – the constant stream of charged particles given off by the sun at high speeds.

Index

Visit abdokids.com
to access crafts, games,
videos, and more!

Use Abdo Kids code
SNK9098
or scan this QR code!